AN IMAGINATION LIBRARY SERIES

Colors of the Sea

CORAL REEF PARTNERS

Eric Ethan and Marie Bearanger

Gareth Stevens Publishing

MILWAUKEE

For a free color catalog describing Gareth Stevens Publishing's list of high-quality books and multimedia programs, call 1-800-542-2595 (USA) or 1-800-461-9120 (Canada). Gareth Stevens Publishing's Fax: (414) 225-0377. See our catalog, too, on the World Wide Web: http://gsinc.com

Library of Congress Cataloging-in-Publication Data

Ethan, Eric.
    Coral reef partners / Eric Ethan and Marie Bearanger.
      p. cm. — (Colors of the sea)
    Includes index.
    Summary: Describes some of the symbiotic relationships that help
preserve the fragile coral reef ecosystem.
    ISBN 0-8368-1740-0 (lib. bdg.)
    1. Coral reef ecology—Juvenile literature. 2. Symbiosis—Juvenile
literature. [1. Coral reef ecology. 2. Symbiosis. 3. Ecology.]
I. Bearanger, Marie. II. Title. III. Series: Ethan, Eric. Colors
of the sea.
QH541.5.C7E84   1997
577.7'89—dc21   96-47565

First published in North America in 1997 by
Gareth Stevens Publishing
1555 North RiverCenter Drive, Suite 201
Milwaukee, WI 53212 USA

This edition © 1997 by Gareth Stevens, Inc. Adapted from *Colors of the Sea* © 1992 by Elliott & Clark Publishing, Inc., Washington, D.C. Text by Owen Andrews. Photographs © 1992 by W. Gregory Brown. Additional end matter © 1997 by Gareth Stevens, Inc.

Text: Eric Ethan, Marie Bearanger
Page layout: Eric Ethan, Helene Feider
Cover design: Helene Feider
Series design: Shari Tikus

The publisher wishes to acknowledge the encouragement and support of Glen Fitzgerald.

Printed in the United States of America

1 2 3 4 5 6 7 8 9 01 00 99 98 97

TABLE OF CONTENTS

WHAT IS SYMBIOSIS?

Most types of fish and other sea creatures are **predators**. This means they hunt and feed on other sea life to stay alive. But some sea creatures have become partners instead. Sometimes this partnership is one-sided. In other cases, both of the animals help each other.

When two animals behave in ways that benefit both of them, the process is called **symbiosis**. Animals do not help each other because they like each other. They do it to survive.

A cleaner shrimp, *Periclimenes pedersoni*, and a cleaning goby, *Gobiosoma genie*, pull **parasites** from a Caribbean graysby, *Cephalopholis cruentatus*.

WHAT ARE COMMENSAL PARTNERSHIPS?

Some partnerships between two sea creatures help just one of them most of the time. The other creature is not harmed, but it does not get as much as it gives. These are called **commensal** partnerships.

An example of this is the anemonefish, which gets its name from hiding in the **tentacles** of sea creatures called sea anemones. Sea anemones have poisonous stingers on their tentacles that they use to hunt prey. But the anemonefish is **immune** to sea anemone stingers. This gives the anemonefish a safe place to hide and from which to hunt. Sea anemones do not get anything from this relationship, but they are not harmed by it either.

A white-striped anemonefish, *Amphiprion perideraion*, hides in the tentacles of sea anemones.

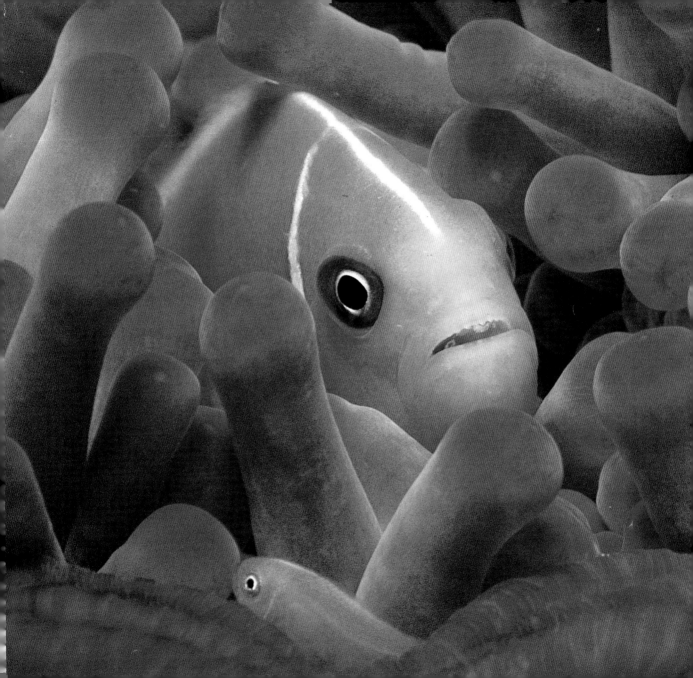

WHAT ARE MUTUAL PARTNERSHIPS?

Besides being symbiotic, partnerships between sea creatures when both creatures benefit are **mutual** partnerships.

For example, anemone shrimp or cleaner shrimp also hide in the tentacles of sea anemones for protection and are immune to the stinging tentacles. But, unlike anemonefish, cleaner shrimp help the sea anemones. They keep the anemones healthy by pulling off pieces of dead tissue from them. In turn, this dead tissue is a source of food for the cleaner shrimp.

An anemone shrimp, or cleaner shrimp, *Periclimenes yucatanicus*, is shown among the tentacles of a giant sea anemone.

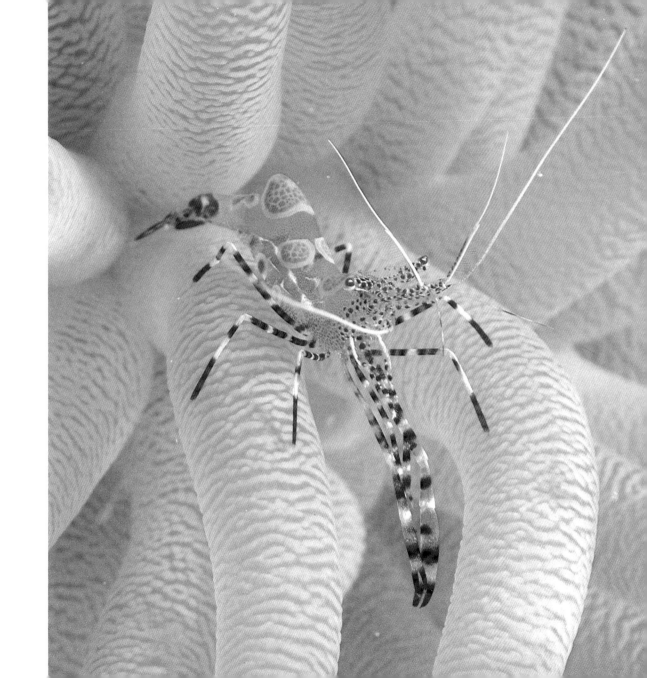

WHAT ARE PARASITES?

Parasites are not welcome partners. They attach themselves to their "host" partners and feed on them to survive. The hosts are able to survive as well.

Small crustaceans, or hard-shelled water animals, called copepods and isopods, are common parasites. They attach themselves to fish. Special cleaner fish pull these unwelcome parasites from their hosts. Cleaner fish have a mutual partnership with the host fish.

A cleaner shrimp, *Periclimenes*, pulls parasites from a coral trout, *Cephalopholis miniatus*.

HOW DOES ALGAE HELP CORAL?

The most important mutual partnership on coral reefs is between coral **polyps** and **algae**. Coral polyps are tiny animals that live inside their limestone shells or skeletons that make up coral reefs. Algae called zooxanthellae grow on most coral polyps.

The polyps feed on **plankton**, giving off substances called nitrogen and **phosphates**. Algae need these substances to live. The algae, in turn, produce sugars that are an important food source for most corals.

The sawcheek cardinalfish, *Apogon quadrisquamatus*, lives among the stinging tentacles of giant sea anemones.

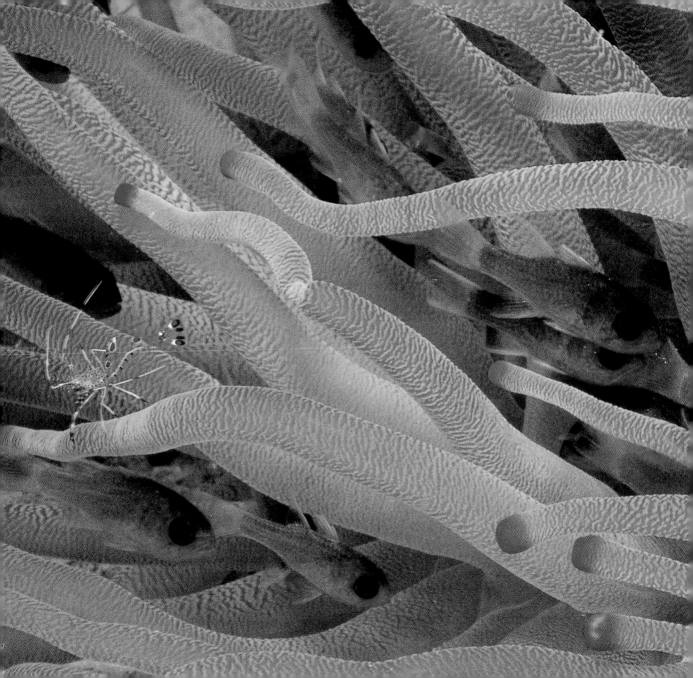

Zooxanthellae help coral in another way. The algae keep phosphates that polyps give off from building up. This helps polyps add to their hard limestone homes. Marine scientists are not sure how this process works, but they do know that coral reefs build up much faster as a result of the partnership between polyps and algae.

Besides providing algae with food, coral polyps also give algae a place to grow in the warm, shallow waters where they do best. This is an example of symbiosis or a mutual partnership.

Beautiful tube sponges and vase sponges are pictured near algae-covered coral reefs.

FINDING PARTNERS

Partnerships near the coral reefs are not always easy for marine scientists to observe. For example, the partnership between algae and coral takes place at a level too small for the human eye to see. It is easier to observe partnerships that exist between the larger sea creatures.

Scientists have studied how partners in the sea find each other. The cleaning wrasses and cleaning gobies have bright stripes on their sides. These stripes are called **guild signs**. The stripes help larger predator fish recognize the cleaning fish, allowing the cleaners to get close enough to the larger fish to pull parasites from them.

A cleaning wrasse approaches a golden damselfish, *Amblyglyphidon aureus*.

Cleaning wrasses and cleaning shrimp often stay in certain areas around reefs. Other fish seem to know they can have parasites removed if they go to these areas.

When a fish is being cleaned of parasites, it remains very still, almost seeming to enjoy it. The larger fish could easily catch and eat the cleaning wrasses and shrimp, but they almost never do.

These double-bar anemonefish, *Amphiprion chrysopterus*, take shelter in the tentacles of a small sea anemone.

FRAGILE ECOSYSTEM OF THE REEF

Coral reefs have existed for thousands of years. During that time, the reefs have survived many threats.

When coral is out of the water, it seems very hard. But, in fact, coral reefs are very fragile and can be easily damaged. During the last century, humans have damaged this fragile **ecosystem** a great deal.

One threat comes from people taking coral for souvenirs and jewelry. Rare kinds of coral have been taken from the ocean so often that they have almost completely disappeared. There are now limits on the amount of certain corals that humans can take from the reefs.

Humans, too, must become responsible partners with coral reefs in order for the reefs to survive.

Coral reefs have also been damaged because of human efforts to build harbors and channels into them. It takes nature many centuries to replace what machinery can remove in just a few hours.

Humans are also at the root of the greatest threat of all to the coral reefs — **pollution**. Some pollutants hurt the coral polyps and the algae they rely on. Other pollutants kill the creatures that make up the plankton that coral feeds on.

Humans must learn to value the unique ecosystem of the coral reef. In return, humans benefit from the beauty and wonder of the magnificent reefs.

GLOSSARY

algae (AL-jee) — Water plants that are food for many sea creatures.

commensal (ko-MEN-sel) — A relationship between two different kinds of animals where only one of them benefits.

ecosystem (EH-ko-sis-tum) — A community of plants and animals.

guild sign (gild sine) — A marking on certain animals that keeps them safe from predators.

immune (im-MYOON) — Protected against something that is harmful.

mutual (MYOO-choo-al) — A relationship between two different kinds of animals where both of them benefit.

parasite (PARE-ah-site) — A living being that attaches itself to and feeds off another living being.

phosphates (FOS-fates) — Important chemical foods for plants.

plankton (PLANK-ton) — Tiny plants and animals floating in the water that are a source of food for sea creatures.

pollution (pah-LOO-shun) — The toxic wastes or poisons in the air, land, or water.

polyp (PAH-lip) — A small animal that lives in the water; it has a tube-shaped body, a mouth surrounded by tentacles, and a limestone shell or skeleton.

predator (PRED-a-ter) — An animal that lives by eating other animals.

symbiosis (sim-be-OH-sis) — The relationship between two different types of animals that results in benefit for both.

tentacle (TENT-ah-cuhl) — A flexible, tubelike arm of a sea creature that is used for collecting food, holding, moving, or stinging.

WEB SITES

http://www.blacktop.com/coralforest/

http://planet-hawaii.com/sos/coralreef.html

PLACES TO WRITE

The Cousteau Society, Inc.
870 Greenbrier Circle, Suite 402
Chesapeake, VA 23320

Environmental Protection Agency
Oceans and Coastal Protection Division
401 M Street SW
Washington, D.C. 20460

Greenpeace (USA)
1436 U Street NW
Washington, D.C. 20009

Greenpeace (Canada)
2623 West Fourth Avenue
Vancouver, British Columbia V6K 1P8

Greenpeace Foundation
185 Spadina Avenue, Sixth Floor
Toronto, Ontario M5T 2C6

Center for Marine Conservation
1725 DeSales Street, Suite 500
Washington, D.C. 20036

National Geographic Society
17th and M Streets NW
Washington, D.C. 20036

INDEX